IN THIS GUIDE

STONEHENGE AOTEAROA

Stonehenge Aotearoa is unique in New Zealand and internationally as a place of science and wonder. Inspired by the original 4000-year-old prehistoric monument on England's Salisbury Plains, members of the Phoenix Astronomical Society set out to create a modern version – Stonehenge as it might have been had it been built in the Southern Hemisphere, at this particular latitude and longitude.

What would such a structure reveal about the movement of the sun, the stars and the planets? What could be learnt about the mythologies of ancient peoples? And what insights ... zing ... l and ... ght the ... present day?

Stonehenge Aotearoa shows in a captivating way the links between astronomy, archaeology, anthropology, history, the social sciences and Maoritanga – Maori customs and beliefs. The aim of this guide is to encourage everyone, irrespective of background, age or ethnicity, to become inspired by science – particularly astronomy – and to gain a greater understanding of the physical universe around us.

TAKE A TOUR OF STONEHENGE AOTEAROA

The vision for Stonehenge Aotearoa was to create a practical open-sky observatory inspired by, and built on a similar scale to, the famous Stonehenge in England. It is not an exact replica of this mysterious ancient monument but a modern interpretation, inspired by the many stone circles and henges scattered around the globe.

Stonehenge Aotearoa combines modern scientific knowledge with Celtic and Babylonian astronomy, Polynesian navigation, and Maori starlore and maramataka (the calendars of time and seasons). As it is a natural observatory, it enables the study of Ngaa huri o te waatau – The turning of the seasons – and Ngaa huri o ngaa whetu – The turning of the stars. Visitors can not only become aware of the wonders of the cosmos but also, regardless of the culture from which they come, reclaim a little of the knowledge of their ancestors.

THE LOCATION

Stonehenge Aotearoa is designed specifically for its location in the Wairarapa region of New Zealand's North Island. In the centre of the henge is a bronze compass rose marked with the cardinal points – the compass bearings of north, east, south and west.

Stonehenge Aotearoa coordinates are:

LATITUDE: 41° 06′ 04″.8 South

LONGITUDE: 175° 34′ 24″ East,
or 11h 42m 17.6s ahead
of coordinated universal time

ELEVATION: 94 metres above
mean sea level.

THE STRUCTURE

The basic henge consists of **24 upright pillars**, connected by **lintels** to form a circular structure 30 metres in diameter and about 4 metres high. This structure is similar to the sarsen circle of the original Stonehenge and has the same diameter.

Enter via the **causeway**, which runs due west to the centre. As you approach the henge there are a fountain and a line of standing stones to either side of the causeway.

Near the centre of the henge stands a 5-metre-high **obelisk**. Running out from this, along the meridian (see page 21), is a 10-metre-long tiled area called an **analemma**.

Ten metres outside the circle of the henge stand **six heel stones** of varying heights.

If you stand at the centre of the henge, the pillars and lintels appear to form doorways. These frame the rising and setting points of the sun, the moon, and bright stars that are either important seasonal markers or navigational beacons. The obelisk, analemma and heel stones also provide ways to track the movement of the sun and the moon over the seasons.

Seen from the centre, the pillars and lintels form windows or doorways along the horizon.

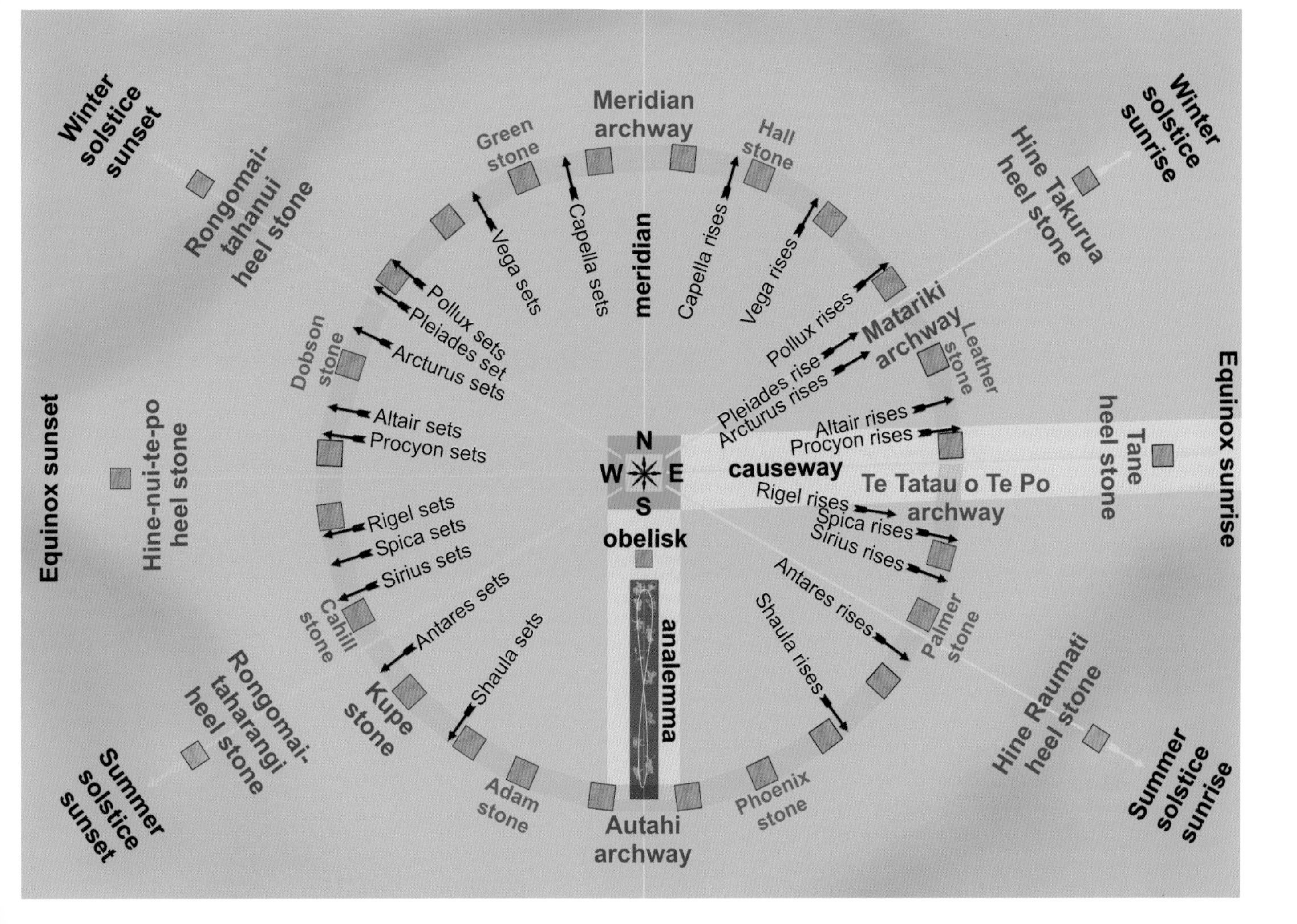

Winter solstice sunset
Rongomai-tahanui heel stone
Green stone
Meridian archway
Hall stone
Hine Takurua heel stone
Winter solstice sunrise
Vega sets
Capella sets
meridian
Capella rises
Vega rises
Pollux sets
Pleiades set
Arcturus sets
Dobson stone
Pollux rises
Pleiades rise
Arcturus rises
Matariki archway
Leather stone
Equinox sunset
Hine-nui-te-po heel stone
Altair sets
Procyon sets
N
W
E
S
Altair rises
Procyon rises
causeway
Te Tatau o Te Po archway
Tane heel stone
Equinox sunrise
Rigel sets
Spica sets
Sirius sets
obelisk
Rigel rises
Spica rises
Sirius rises
Cahill stone
Antares sets
Shaula sets
analemma
Antares rises
Shaula rises
Palmer stone
Kupe stone
Rongomai-taharangi heel stone
Hine Raumati heel stone
Summer solstice sunset
Adam stone
Autahi archway
Phoenix stone
Summer solstice sunrise

THE EARTH'S TILT & THE SEASONS

As the Earth orbits around the sun, the north and south poles are alternately tilted towards the sun. The sun's altitude therefore increases and decreases during the year, producing seasons.

This diagram shows the Earth at four points in its orbit around the sun. In December the south is tilted towards the sun, while the north is tilted away – it is summer in the Southern Hemisphere and winter in the Northern Hemisphere.

Six months later, in June, the situation is reversed and the Southern Hemisphere is tilted away from the sun – it is winter in the Southern Hemisphere and summer in the Northern.

The points that mark midwinter and midsummer are called the solstices. These occur every year, depending upon your time zone, on either June 21 or 22 and December 21 or 22.

There are two midway points, one in March and another in September, when neither hemisphere is tilted towards the sun.

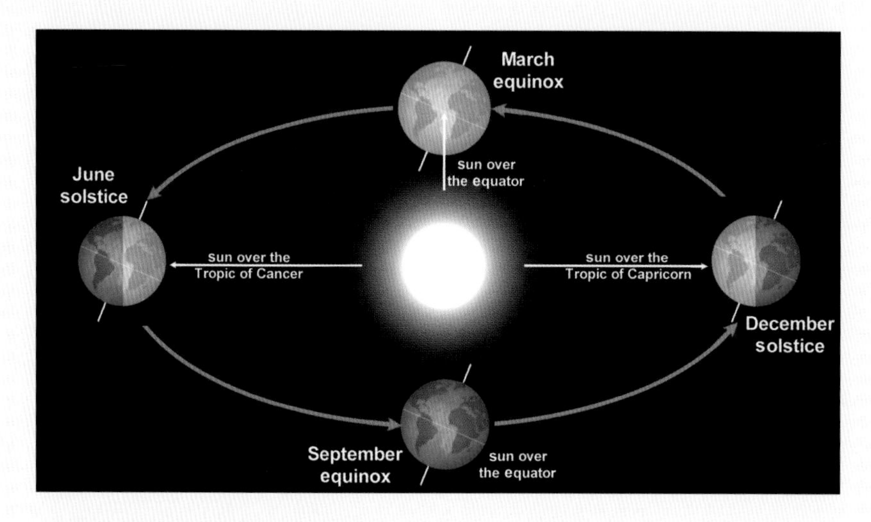

These are the equinoxes. March is the Southern Hemisphere autumn and Northern Hemisphere spring; September is the Southern Hemisphere spring, and Northern Hemisphere autumn. The equinoxes occur on March 20 or 21 and September 22 or 23.

THE HEEL STONES

The western heel stones, with the winter solstice stone in the foreground.

The heel stones are positioned to mark the rising and setting points of the sun at the midsummer and midwinter solstices, and the autumn and spring equinoxes. The rising and setting positions of a given star remain the same throughout the year, but those of the sun, moon, planets and other objects within our solar system vary.

The heel stones and the sun

The rising and setting position of the sun is observed to move steadily southward from midwinter (23° 26′ N) to midsummer (23° 26′ S), and then steadily northward from midsummer through the equinox (0° 00′) to midwinter.

In addition, the rate of movement of the sun along the horizon – the change in the place where it rises and sets – varies throughout the year. Near a solstice the movement from day to day is slow, but between solstices, particularly at an equinox, it is rapid.

Some Maori tribes have a story that illustrates this well. Te Ra, the sun, has two wives, Hine-Raumati and Hine-Takurua, bright stars that mark the seasons. Over the year he moves from one wife to the other. At the winter solstice he rises with Hine-Takurua, the winter maid (the 'frost star' Takurua, or Sirius). Six months later he rises with the summer maid, Hine-Raumati (Rehua, or Antares).

At dawn over the year, you can watch Te Ra move from one wife to the other. At a solstice, as he reluctantly leaves the comfort of a wife, he moves slowly. But at an equinox, when he is between the wives and both can see him – both stars are in the sky – he moves very quickly.

The Maori legend of the first house

According to Maori legend, the first house or Wharekura – the house of learning in which we all live – was formed by the separation of the first parents: Ranginui, the sky father and Papatuanuku, the earth mother. Papa is the base and Rangi the roof of the house.

In this symbolic representation of the Maori story of the Wharekura, the first house, Te Ra, the sun, rises over the Tane stone (centre) at the equinox. Tane, the great god of the forests and people, is also in Maori legend a personification of the rising sun. The heel stones either side are his wives, Hine-Takurua (left) and Hine-Raumati. The sun rises over Hine-Raumati in the south-east on midsummer's day. On midwinter's day the sun rises in the north-east over Hine-Takurua.

The stars are shown in their position at the beginning of the Maori New Year, heralded by the first rising of Matariki, the Pleiades star cluster, over the stone of Hine-Takurua. The archway leading to this stone is named Matariki.

ABOVE: *In Maori astronomy, bright stars along Te Ikaroa, the Milky Way, mark the seasons. The Milky Way is also known as the Parent Way. The two solstice heel stones in the west represent Rongomai-taharangi and Rongomai-tahanui, the great guardians of the seasons on the Parent Way. This photo shows the sun over the Rongomai-taharangi heel stone at Stonehenge Aotearoa.*

Hine-nui-te-po stone

The post at the back of the first house of Maori legend is that of Hine-nui-te-po, the great goddess of death. It is aligned due west, where the sun sets at an equinox. At Stonehenge Aotearoa this post is represented by the equinox heel stone in the west. Just after sunset in late May, an upright line of four bright stars rests directly above this stone. These stars are the post in the sky and mark the death, or end, of the Maori year.

The great goddess of death, Hine-nui-te-po, is pictured above her stone at sunset. In May, at the end of the Maori year, an upright line of four bright stars rests directly above this stone. To Maori they are the post at the back of the wharekura, the first house.

Tane stone

The post at the front of the first house due east, is the Tane post. At Stonehenge Aotearoa this is represented by the Tane stone. This marks where the sun rises at an equinox.

Te Tatau o Te Po

The doors of the house are the four winds, or four cardinal points. The main entrance, Te Tatau o Te Po (The doorway of the night), lies where the sky meets the sea. This is also due east, where the sun rises at an equinox. At Stonehenge Aotearoa, the east-aligned archway of stones at the main entrance is named Te Tatau o Te Po.

THE POSITION OF THE EQUINOX STONES

If Stonehenge Aotearoa were located on a flat plain, the two equinox stones – Tane and Hine-nui-te-po – would be directly opposite each other, one due east and the other due west. However, because its location is inland, the horizon is uneven, and the rising and setting positions of the sun, and hence the positions of the stones, are slightly offset from the cardinal points of the compass.

The heel stones and the moon

When the moon is full, the places along the horizon where it rises and sets, and its altitude when it crosses the meridian, will be the seasonal opposite of the sun's.

At the midsummer solstice, when the sun is 23.5 degrees south of the celestial equator and rises over summer solstice heel stone Hine-Raumati, the full moon will be on the opposite side of the sky, approximately 23.5 degrees north of the celestial equator, and will rise close to the winter solstice heel stone Hine-Takurua.

The sun will rise in the south-east and be high in our midday sky. The moon will rise in the north-east (the sun's winter rising position) and be low in the sky when it crosses the meridian.

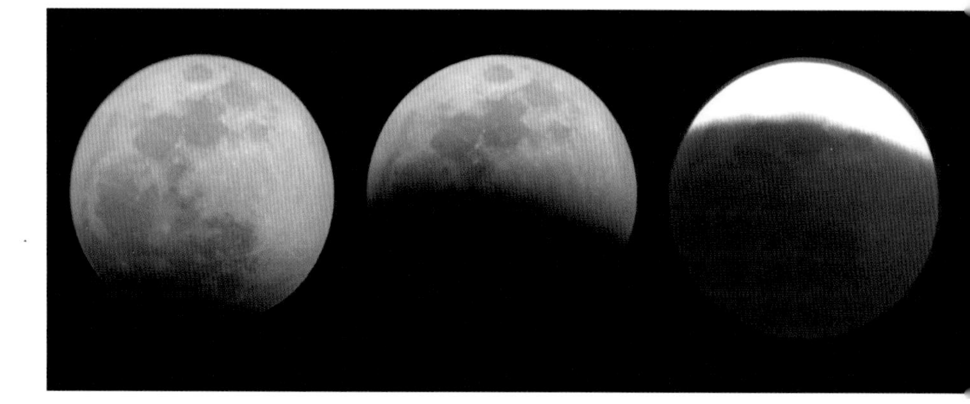

At the winter solstice the opposite occurs – the sun is low in the sky and the moon is high. Only at the equinoxes do the sun and full moon have approximately the same altitude when they cross the meridian, and both then rise due east and set due west.

Note that this system of opposites applies only with a full moon. During the moon's monthly cycle it travels right around the ecliptic – the path of the sun – and through all the constellations of the zodiac. However, you can still determine the location of the sun by the position and phase of the moon.

For example, at first quarter the moon is at right angles to the sun and 90 degrees along the ecliptic ahead of the sun. When the sun reaches the southern winter solstice, the first quarter moon will be at the position of the southern spring equinox.

Eclipse of the moon, 16 July 2001.

ECLIPSES

There is some evidence that structures at the original Stonehenge were used to predict eclipses. An eclipse can occur only when the sun, moon and Earth are directly in a line. A solar eclipse can therefore take place only when the moon is new, and lying between the Earth and the sun — that is, it is on the part of its orbit that crosses the path of the sun.

To complicate matters, the moon's orbit is tilted 5^0 8' to the celestial path of the sun, and as a result its rising and setting positions vary slightly from month to month. If they didn't, there would be a solar eclipse every month.

The shadow cast by the moon is little more than a thousand kilometres wide when it reaches the Earth. To see an eclipse you have to be in the shadow's path. This doesn't happen very often at any one locality; however there is a pattern, or sequence, of eclipses that repeats itself over a period of 18.5 years. By watching the rising positions of the sun and moon and counting the number of new moons, you can predict when a solar eclipse is likely to occur.

A lunar eclipse occurs when the Earth moves between the sun and the moon. Thus, lunar eclipses occur only when there's a full moon. They are more frequent than solar eclipses because the Earth is larger than the moon and therefore casts a larger shadow.

During a lunar eclipse the moon turns a red or copper colour. This illumination comes from sunlight which is refracted, or bent, on to the moon by the Earth's atmosphere. The refraction affects red light more than other colours (or wavelengths) in the spectrum, hence the red hue.

THE STONE CIRCLE

On a clear dark moonless night, well away from city lights, the Milky Way can be seen as a mottled band of glowing light, a great highway of bright stars that arches across the heavens from one horizon to another. Throughout the night and through the seasons, this great highway slowly turns, changing its orientation across the sky.

The rising point, east of the meridian, steadily moves northward along the horizon. The setting point, west of the meridian, moves southward. Thus from the centre of the circle of Stonehenge Aotearoa, a procession of bright stars can, with the passage of time, be seen to rise in one window after another in a northerly direction. West of the meridian the procession of setting stars can be observed in a series of windows moving southward.

Through the course of the year, the Milky Way moves from one window to the next. Colours that cannot be seen by the naked eye are captured by the camera in this dramatic photograph.

The red glow above the summer solstice heel stone is the Lagoon nebula. The bright reddish star Antares, the heart of the Scorpion, appears above the left pillar; the tail curls to the right and left again above the central lintel. The two bright stars at the tail are the Scorpion's stinger – to Maori, the barbs of the fish-hook of Maui.

CONSTELLATIONS

In the European and west Asian cultures that arose from ancient Mesopotamia, Scorpio (above) is seen as a scorpion, but in China it is a dragon and in Hawaii the fish-hook of the legendary demi-god Maui. The bright red star Antares, the heart of the Scorpion, rises and sets close to the summer solstice stones. Shortly after the Scorpion has risen it can be seen lying along the tops of the south-eastern lintels. In this photo the Scorpion's heart can be seen centre left, with the tail curling right and back again to the left.

Today's sky is mapped by dividing it into 88 specific regions called constellations.

Stars in a constellation often have no physical relationship with each other: the patterns are simply perceived by people on Earth. In addition, a constellation's pattern in the sky usually bears no resemblance to the figure – such as a water carrier or goat – commonly used to represent it. This is because most constellations were named in ancient times for their seasonal importance, rather than their shape in the sky.

Two of the most notable constellations are Orion, the Hunter, and Scorpius, the Scorpion, both composed of bright stars in the Milky Way. They are almost opposite each other in the sky, and have traditionally been used around the world as seasonal markers.

In Aotearoa-New Zealand, Orion dominates the summer evening sky, first appearing in the midwinter dawn twilight over the Tane stone. Scorpius dominates the winter evening sky, first appearing at midsummer over the stone of Hine-Raumati. The two are seen together only in spring and autumn, when one sets in the west as the other rises in the east.

In northern cultures the constellation of Orion (above) has always been depicted in the figure of a hunter, while to Maori it is sometimes Tautoru, the bird-snarer. The Belt of Orion, a line of three bright stars, rises due east, and sets due west over the equinox heel stones.

HERALD STARS

The rising of certain stars has traditionally been seen as ushering in seasonal events. For example, the rising of Matariki, the Pleiades star cluster, in June's dawn twilight traditionally heralded the start of a new year for many Maori iwi, or tribes. For others the year began with the rising of Puanga, the bright blue-white star in Orion known as Rigel. At Stonehenge Aotearoa, Matariki rises above the winter solstice stone Hine-Takurua, while Puanga rises in the archway of Te Tatau o Te Po.

To many Maori, stars were thought to be food-bringers, so certain seasons and periods were named after them. For example, the month of Whakaahu, the time of new growth, is named after the star commonly known as Pollux, which first rises in the spring dawn. The month of Poutu-te-rangi, the time of the kumara harvest, is named for Altair, which rises in the dawn twilight at the very time the kumara harvest should be inspected for maturity.

NAVIGATIONAL STARS

Most of the stars used for navigation by seafarers lie in the Milky Way. These stars form a navigational pathway from the Pacific islands around the equator to Aotearoa-New Zealand.

According to legend, the great Polynesian chief Kupe instructed voyagers to Aotearoa: 'Lay it [the bow of the canoe] to the right of the sun in the month when Venus is in the heavens [October], but go in the summer, at the height of the year [November/December].'

What he meant was that bearings should be taken when Venus was visible, but that the voyage should not be undertaken until later in the year, when suitable south-west winds would be created by tropical anticyclones. Seen from the centre of Stonehenge Aotearoa, the Kupe stone is aligned with the direction of the migrations from the northern islands (the traditional Hawaiiki) to Aotearoa.

Stars in the fish-hook of Maui (part of the constellation of Scorpius) mark the direction to be taken when sailing to Aotearoa (New Zealand) from the legendary homeland of Hawaiiki. At certain times of the year, these stars pass below the horizon behind the Kupe stone.

In the foreground, superimposed, is the image of a statue on Wellington's waterfront of, from left, Hine-te-aparangi, Kupe and Ngahue. It was Hine-te-aparangi who first sighted Aotearoa. She called out,
'He ao, he ao! He aotea! He aotearoa' – 'A cloud, a cloud! A white cloud! A long white cloud.'

The bright stars used by the ancient mariners still give guidance to sailors today. In the Southern Hemisphere the most celebrated are the four stars of the Southern Cross, and Canopus and Achernar. These stars are used to locate the south celestial pole and the south cardinal point, and to determine latitude.

From Aotearoa all of these stars are circumpolar, meaning they never set. Their orientation in the sky will change from night to night and from season to season, but like hands on a clock they move in great circles around the celestial pole; at Stonehenge Aotearoa they move around the pinnacle of the obelisk. Because of their clock-like nature they are also used as markers of seasons.

Any bright star which is so far north that it is never far from the horizon makes an excellent marker for the north cardinal point. When the star reaches its highest altitude, due north will be on the horizon directly below. A star which does this is Capella. As Canopus approaches the zenith – the point directly overhead – Capella is due north.

Another important northern star is Vega, which to Maori is the sacred star Whanui, the great chief who sails from the Maori ancestral homeland. When Canopus is at its lowest point and due south, Vega is due north.

ZENITH STARS

Certain stars, known as zenith stars, are used to determine the latitude of specific islands. Shaula, the sting in the tail of the Scorpius constellation, is the zenith star for Aotearoa, and is known to Maori as Potiki. Sailors travelling from the north knew that when this star was directly overhead they had reached the latitude for Aotearoa. They would then turn either east or west, depending on their path, to reach land.

This composite photograph shows the southern night sky on a winter evening in the Wairarapa. Just above the horizon is Canopus, the second brightest star in the sky. To Maori Canopus is Autahi, the high chief of stars who hauls up the great anchor Te Punga, the Southern Cross, and sets in motion the cycle of the seasons.

Above and to the left of Canopus is the Large Magellanic Cloud. The bright glowing patch at the centre is the Eta Carinae nebula. Above this, lying on its side, is the Southern Cross, with the dark Coalsack nebula. Above the Cross are the two bright 'Pointer Stars', Alpha Centauri (uppermost) and Agena.

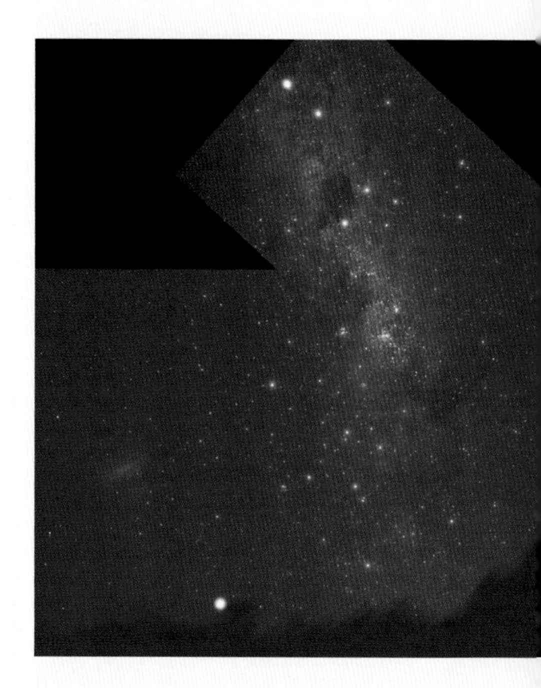

The bright northern star Whanui (Vega) passes through the Meridian archway at Stonehenge Aotearoa. Just above the lintel lies the bright star Poutu-te-rangi (Altair).

THE OBELISK

The obelisk at Stonehenge Aotearoa stands on the meridian, an imaginary line running due north to due south through the centre of the henge. The word obelisk comes from the Greek *obeliskos*, meaning a prong for roasting. In ancient Egypt, an obelisk was a needle of stone placed over a temple. Dedicated to Ra, the sun god, it was believed to be a protective petrified solar ray. Obelisks were also erected in ancient Assyria in the twelfth century BCE and Babylonia in the ninth.

An observer standing near the centre of Stonehenge Aotearoa can sight the south celestial pole, the point around which the entire sky appears to rotate, by looking along the slope of the top cap of the obelisk.

The obelisk and analemma at Stonehenge Aotearoa.

Alternatively it can be sighted through the hole in the body of the obelisk: this allows a person of any height to find the south celestial pole. Simply walk towards or away from the obelisk until the hole looks like a circle. You are then looking at the south celestial pole and facing due south.

If a camera is lined up on the hole in this way and a long exposure photograph taken, as the Earth rotates the stars draw luminous trails in a clockwise direction around the hole. At Stonehenge Aotearoa, the south celestial pole is 41° 06′ 04″.8 above mean sea level. This is the exact latitude of the henge.

As the Earth rotates the stars appear to move in circles around the south celestial pole, producing star trails in a long exposure photograph taken from the centre of the henge.

THE MERIDIAN

The meridian, an imaginary line in the sky similar to the lines of longitude on Earth, runs directly overhead from north to south.

In the Southern Hemisphere the meridian runs from the south celestial pole, through the zenith (directly overhead), across the celestial equator, and onwards below the horizon to the north celestial pole.

When the sun, the moon or a star crosses the meridian, it reaches its highest point in the sky for that latitude.

THE ANALEMMA

An analemma is the path traced by noting the daily position of the sun at the same time each day.

This path is shaped like an elongated figure eight. An analemma may be calculated by recording the sun's position in the sky each day at the same time, or by recording the position of a shadow cast by the sun.

The analemma at Stonehenge Aotearoa is defined by the shadow cast by the 5-metre-tall obelisk. The length of this shadow varies over the year by 8.7 metres. When the sun is furthest south, at the midsummer solstice, the shadow of the midday sun will be at its shortest, 1.55 metres. At the midwinter solstice the shadow will be at its longest, 10.25 metres.

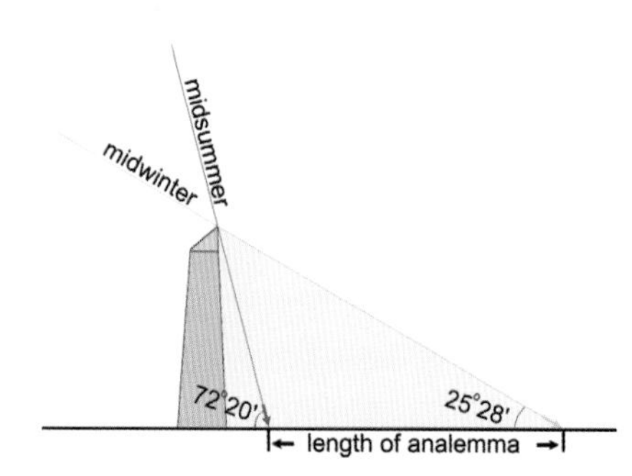

This diagram shows, for the latitude of Stonehenge Aotearoa, the difference in the altitude of the sun and the length of shadow cast by the obelisk at noon at midsummer and midwinter.

The analemma is set into a tile mosaic running along the meridian south of the obelisk. The design is graduated so the shadow cast by the obelisk points to the date with an accuracy of about two or three days. It incorporates the constellations of the zodiac, and the location of important stars close to the path of the sun.

The analemma and the sun

The analemma shows how the sun's position, as seen from the Earth, changes through the year relative to the background stars. This change in position is due to the Earth's orbit around the sun.

If the sun crossed the meridian at the same time on the clock each day, the point of the shadow cast by the obelisk would draw a straight line from north to south. However, the sun sometimes runs ahead of clock time, and sometimes behind. Hence the shadow changes each day in both length and direction, generating a figure eight.

The length of a day varies slightly throughout the year – a day is not always exactly 24 hours of clock time. For time-keeping purposes astronomers use the concept of a 'mean sun'. This averages times out over the year so that a day is always exactly 24 hours.

New Zealand Standard Time (NZST) is based on the time the 'mean sun' would cross the 180° longitude meridian, which is exactly 12 hours ahead of the time it would cross the longitude zero meridian at the Royal Observatory in Greenwich, England.

The longitude of Stonehenge Aotearoa is 175° 34′ 24″ East, or in units of time 11h 42m 17.6s. New Zealand Standard Time is based on a longitude of 180° 00′ 00″, or 12h 00m 00s. This means the 'mean sun' crosses the meridian at Stonehenge Aotearoa 0h 17m 42.4s after it crosses the 180° meridian – that is, at nearly 12:18pm each day.

Therefore the analemma at Stonehenge Aotearoa can be used as a time check: the point of the shadow cast by the obelisk will each day cross the track of the analemma at 12:18pm NZST, or at 1:18pm when daylight saving is in force.

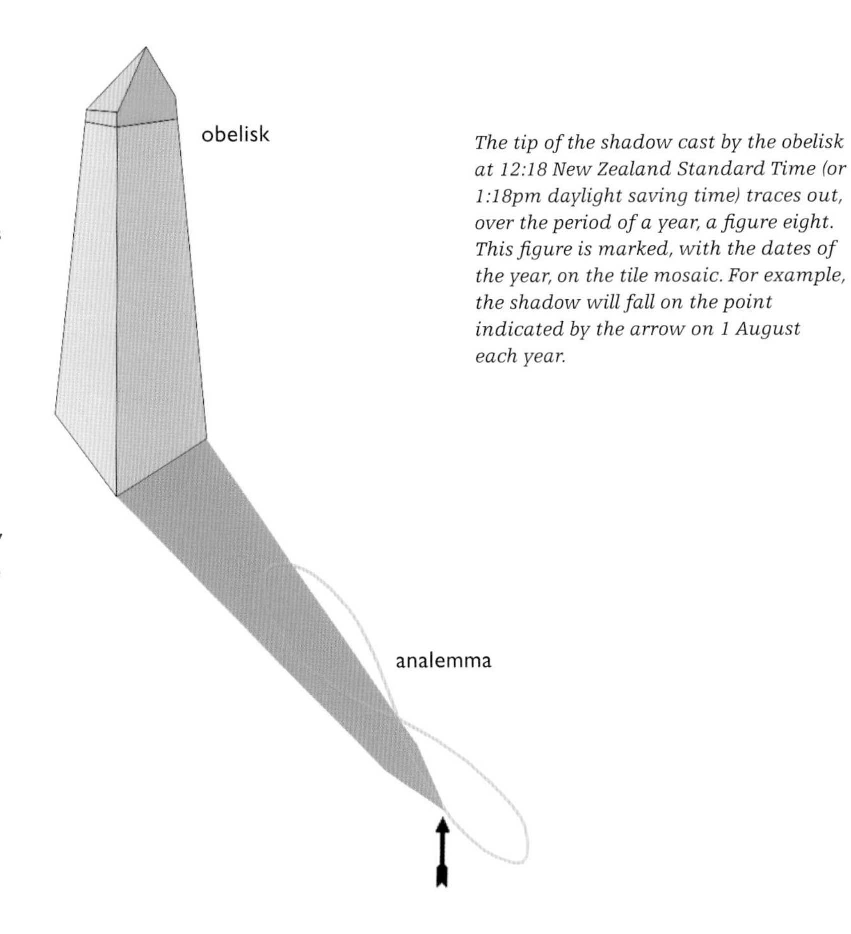

The tip of the shadow cast by the obelisk at 12:18 New Zealand Standard Time (or 1:18pm daylight saving time) traces out, over the period of a year, a figure eight. This figure is marked, with the dates of the year, on the tile mosaic. For example, the shadow will fall on the point indicated by the arrow on 1 August each year.

At certain times the moon approaches its elder brother, the sun, and the two move together for a period … After a time the moon leaves the sun behind … After a space the sun says to the moon, 'Now return to your own place … Pursue your course … In the days that lie before you will return to me.'

Elsdon Best
ASTRONOMICAL KNOWLEDGE OF THE MAORI

The analemma and the moon

The moon, our celestial clock, takes a similar path to the sun, but completes its circuit of the sky along the zodiac in a month instead of a year. The first calendars were built around the cycle of the moon. Hence the reason we have twelve constellations in the zodiac is probably because there are twelve full moons in a lunar calendar year: these give us our 12 months (moonths).

In the solar calendar in use today there are still 12 months in a year, but they are not related to the moon's cycle. In fact, every two or three years the solar calendar gives us 13 full moons in a year: one month will have two full moons. The second of these is called a blue moon.

Maori traditionally used a lunar calendar, in which full moons divided the year into twelve months and a cycle of activities for each season. The 'nights of the moon' (when the moon was visible) were counted from its first appearance in the evening twilight. Islamic countries and Israel still use a lunar calendar.

When the moon is full, it is directly opposite the sun. Each month, as the sun moves from one constellation of the zodiac to the next, so does the location of the full moon. Even though in the daytime you cannot see stars, you can always tell which constellation the sun is in by looking at the full moon at night: it will be in the zodiacal constellation directly opposite. For example, if the full moon is in Pisces the sun will be in Virgo. The moon also casts a shadow on the analemma, indicating the constellation through which it is moving.

The analemma and the stars

The stars along the sun's path form the twelve traditional constellations of the zodiac. At specific times of the year the sun passes in front of each of these constellations – the shadow of the obelisk will point to the zodiac sign through which the sun is passing.

Because of its importance, the constellation of Orion is also included on the tile mosaic. Although Orion is not along the direct path of the sun, it has been used throughout history as a seasonal marker. The obelisk's shadow indicates when the sun is close to Orion.

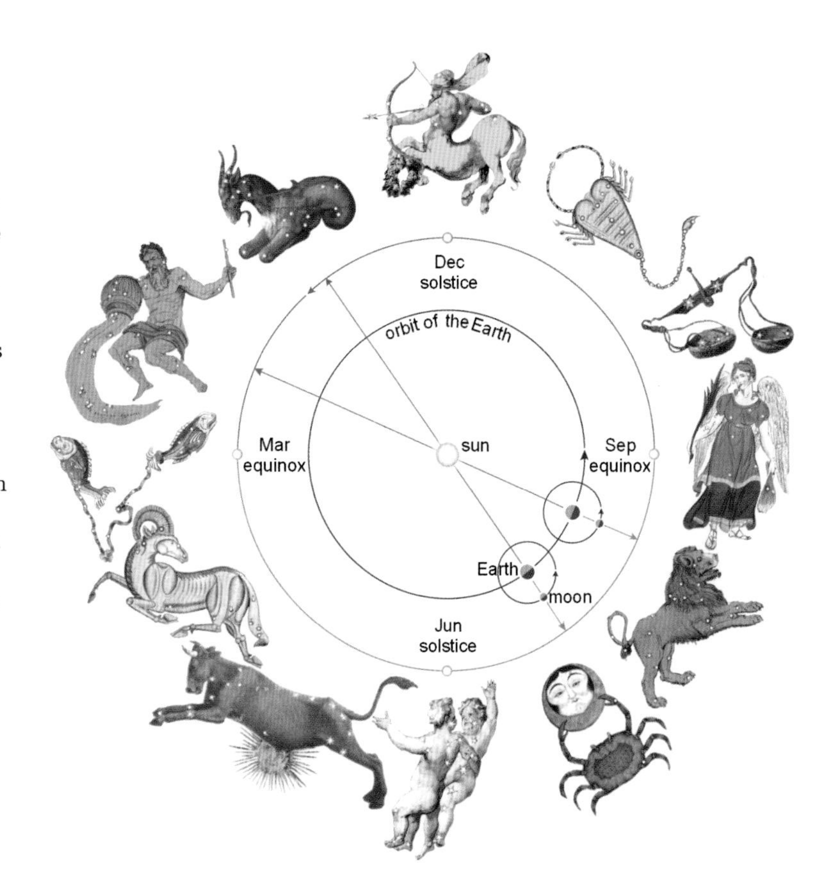

RIGHT: *Because of the orbital motion of the Earth, the sun appears to move against the background stars. Each month the sun and the full moon appear to move from one constellation to the next. For example, in February the sun is in Capricornus, while the full moon is directly opposite in Cancer. A month later the sun has moved into Aquarius and the full moon is in Leo.*

OTHER OBSERVATORIES AT STONEHENGE AOTEAROA

There are two modern observatories at the Stonehenge Aotearoa site, and a third under construction.

A section of the Phoenix Observatory roof rolled back, revealing (left) the Schmidt-Cassegrain, and (centre, rear) the 58cm reflector, while astronomer Richard Hall readies the Peter Read refractor

The **Phoenix Observatory**, opened in July 1999, is designed for recreational observing – to explore the mountains on the moon or the cloud belts of Jupiter, or to peer into the distant realms of jewelled star clusters and glowing nebulae.

It houses four telescopes – the 15cm (6″) Peter Read refractor (on loan from Wellington's Carter Observatory), an automated 25cm (10″) Schmidt-Cassegrain, a 30cm (12″) Dobsonian reflector and a 58cm (23″) Newtonian reflector. Peter Read was for many years New Zealand's most famous astronomer, and popularised the science through his long-running television programme, *The Night Sky*, in the 1960s and '70s.

The **Nankivell Observatory**, opened in February 2005, houses a magnificent 20cm (8″) Schmidt camera with a 15cm (6″) guide refractor built by the late Garry Nankivell, a leading optical craftsman. The camera is designed to reach out into the depths of space and record very faint objects over a wide field.

The **Matariki Research Observatory** will house a 0.6 metre cassegrain telescope designed by New Zealand's foremost optical designer, Norman Rumsey. The innovative design will produce a wide flat field highly suitable for electronic photography. Armed with a good CCD camera, the telescope will resolve objects and structures that a couple of decades ago required the largest telescopes in the world. (A CCD camera, rather than using film to receive light, uses a small piece of special silicon called a charge-coupled device, or CCD.)

The telescope and observatory will be fully automated. The dome and its associated motors and control systems were donated by the United States Navy and were formerly part of its observatory complex at Black Birch in New Zealand's South Island.

The US Naval Observatory at Black Birch near Blenheim, New Zealand being dismantled. Renamed Matariki, it is being rebuilt near Stonehenge Aotearoa.

Over four thousand years ago, when people created that extraordinary sculpture [Stonehenge] in materials that would outlast unimagined generations, they reached out to a vision that added meaning to their lives. If, between that space and the megaliths, we can connect – in whatever way we choose – we show them, and ourselves, a respect that honours our shared humanity. And, from a bunch of old rocks, that's some inspiration.

Mike Pitts
HENGEWORLD

WHY BUILD A NEW STONEHENGE?

Once upon a time a group of keen astronomers dreamed about building a stonehenge in New Zealand...

If this sounds like the beginning of a fairy story, that's because in some ways it is. The building of Stonehenge Aotearoa was a massive undertaking made possible by thousands of hours of unpaid labour by around 150 members of the Phoenix Astronomical Society. The Phoenix Stone is dedicated to these people.

What motivated members of the society to transform a fanciful idea into a reality?

- sharing knowledge of our incredible universe
- teaching astronomy from different cultural perspectives
- teaching the star lore of Maori, the indigenous people of Aotearoa-New Zealand
- providing an educational tool to complement other astronomical facilities on site, and
- creating a uniquely New Zealand astronomical experience.

Why a stone circle?

Many people are put off astronomy because they think it will be too difficult to understand. Stone circles, on the other hand, fascinate and entice a wide range of people. Scottish comedian Billy Connolly is but one of many drawn to dance among such structures (which exist in a number of places in his homeland – see pages 44 and 48)!

Stonehenge Aotearoa's structure enables basic astronomical ideas to be easily understood. Visitors are inspired to continue learning, both about knowledge that was essential to their ancestors' survival, and more recent scientific developments.

How was the project funded?

Having developed the concept, the Phoenix Astronomical Society applied to the Royal Society of New Zealand, which administers the Science and Technology Promotion Fund on behalf of the government, and was successful in obtaining funding. The Royal Society shared the vision of a project that would appeal to, and inspire, people not normally interested in science.

Over one thousand hours of surveying and astronomical calculations went into the design of Stonehenge Aotearoa, work primarily carried out by Phoenix Astronomical Society member Robert Adam, after whom the Adam Stone is named.

The core Stonehenge crew from left to right: Graham Palmer, Chris Cahill, Alan Green, Geoffrey Dobson, Richard Hall (project manager) and Kay Leather (construction team manager). A monolith has been named after each. In addition, the Hall Stone is dedicated to Lesley Hall, who provided valuable support for the project.

DATES

This guide uses culturally neutral terms that have been adopted by the international science community. BC is replaced by BCE (Before the Common Era) and AD by CE (Common Era). A date without letters can be assumed to be CE.

THE ORIGINAL STONEHENGE

For thousands of years people have gazed in awe at the ancient collection of ruins known as Stonehenge on England's Salisbury Plains. Stonehenge is the most famous of hundreds of stone circles found throughout Europe, Africa, Asia and Polynesia. Intriguingly, these mysterious ancient monoliths are part of the history and culture of most peoples on Earth.

While we shall probably never know all the purposes for which Stonehenge was built, there is strong evidence it was used to observe the cycles of the sun, moon and other heavenly bodies. Astronomy is the oldest of the sciences: knowledge of the daily and seasonal changes of these bodies was essential to the survival of early communities, enabling them to predict weather patterns, the best times to plant and harvest crops, animal migrations and other vital information. In this sense structures such as Stonehenge were the first computers, cornerstones to the rise of modern civilisation.

OF KINGS AND CRANKS

When we first announced we intended to build a stonehenge in New Zealand, we were often asked if pagan festivals or Druid ceremonies would be held there. It was even suggested we might be intending to sacrifice virgins. We had to disappoint these people by explaining there was no evidence that human sacrifice was ever carried out at the original Stonehenge. They were even more surprised to learn that the Druid association is pure mythology.

RIGHT: *An artist's impression of the night sky over Stonehenge. It was once believed the structures were created by a wizard, Merlin.*

So who did build Stonehenge and why?

The first mention of Stonehenge is in early twelfth-century writings of Henry of Huntingdon. Henry, an archdeacon at Lincoln Cathedral, was commissioned by Bishop Alexander of Blois to write a history of England.

In it he described Stonehenge as: 'Stanenges, where stones of wonderful size have been erected after the manner of doorways, so that doorway appears to have been raised upon doorway; and no one can conceive how such great stones have been so raised aloft, or why they were put there.'

A few years later a much more elaborate story appeared. In *The History of the Kings of Britain*, Geoffrey of Monmouth, a clergyman and lecturer at Oxford University, described how a fifth-century Saxon king, Hengest, had captured his English counterpart, Vortigern, and murdered several hundred of his lords. After managing to escape to Wales, Vortigern, with the help of a wizard called Merlin, built a great tower on Mount Snowden.

Eventually the rightful British king, Aurelius Ambrosius, returned from exile in Brittany, rallied the British armies and defeated Hengest. He then decided to erect a memorial to the murdered noblemen. Merlin was summoned and suggested he import the Giant's Ring, a huge stone circle on a mountain in Ireland – 'a construction which no man of this age could build, unless he combined great skill with great artistry. The stones are enormous and no one alive is strong enough to move them.'

King Aurelius's brother Utherpendragon (father of the famous King Arthur) duly took 15,000 men to Ireland. After defeating the Irish army, they tried without success to dismantle the Giant's Ring using hawsers, ropes and scaling ladders. Merlin then took down the stones himself, had them loaded on to ships and brought to England, where he re-erected them in a great circle around the graves of the murdered men.

Seen from a distance, Stonehenge appears almost supernatural. Twelfth-century writer Geoffrey of Monmouth promoted the idea that it had been built by a wizard, Merlin. Photo: Diego Meozzi, Stone Pages

Finding the source

This story remained popular for three or four centuries, but by the 1500s, with more people visiting Stonehenge, a more sceptical attitude emerged. In 1534 an Italian-born historian, Polydore Vergil, complained that Geoffrey had extolled the British 'above the noblenesse of the Romains and Macedonians, enhauncinge them with moste impudent lyeing'.

The feeling was growing that Stonehenge had been built by human, not magical, power. Around 1590 a source was identified for the large so-called sarsen stones that comprise the outer circle and the large trilithons in the centre. A local historian named William Lambarde noted there were many such stones on the Marlborough Downs, about 30 kilometres to the north. This has never been challenged, although Sir Christopher Wren, the English architect, astronomer and mathematician, thought the stones, pitched all one way like shot arrows, had been cast up out of volcanoes.

The route by which the stones were taken to the Salisbury Plains has been debated, but modern experiments have shown that with large sleds, and a sufficiently large number of people to pull them, it is possible to transport stones of this weight from one place to another.

A Roman connection?

In the seventeenth century many people speculated about the builders of Stonehenge, although the various theories tended to reflect the interests of the theorisers, rather than objective study. Inigo Jones, architect and

Surveyor of the King's Works, concluded that Stonehenge had been built in the form of a Roman temple. Other people, however, pointed out that the interior didn't have the correct form, being more like an auditorium with a stage. Besides, they said, the Romans would have covered the stones with inscriptions and there were none.

Others saw Stonehenge as the tomb of the warrior queen Boadicea, or the coronation place of ancient Danish kings. One even wrote a book with an elaborate theory involving the ancient Phoenicians. Finally, however, it was recognised that Stonehenge didn't bear much resemblance to any of the things from which it was supposedly derived, and all these theories fell out of favour.

In 1648 John Aubrey, an English antiquarian and writer — now best known as the author of *Brief Lives*, a collection of short biographies — did some fieldwork at Stonehenge and drew a sketch plan of the site. He concluded that Stonehenge was related to other stone circles in the north and west of Britain. As most of these circles were located outside areas of Roman, Saxon and Danish occupation, he argued that they had probably been built as temples by the early native inhabitants of Britain.

Furthermore, he said, it was reasonable to presume that 'the Druids being the most eminent Priests among the Britaines: 'tis odds, but that these ancient Monuments were temples of the Priests of the most eminent order, viz, Druids.' This was the first time Druids, a Celtic priesthood, had been mentioned in connection with Stonehenge.

The Druid fallacy

In the early eighteenth century, Aubrey's fieldwork was continued by William Stukeley, a young physician and friend of the elderly Isaac Newton. From 1721 to 1724 Stukeley spent every summer surveying, drawing and measuring at Stonehenge and in 1740 he published a large volume of the results.

In spite of the excellence of his scientific work, however, Stukeley seems to have been succumbing to what historian Christopher Chippindale would later call 'Druidomania'. While Aubrey had merely speculated that the builders and users of Stonehenge might have been Druids, Stukeley developed the idea into a major obsession, forming a Society of Roman Knights dedicated to saving Roman remains from destruction, and calling himself Chydonax the Druid.

Although the historical evidence on Druids was rather sketchy, there was nothing to suggest they were concerned in any way with stone circles. Undeterred, Stukeley took holy orders, and to give the Druids a fitting Old Testament connection claimed they were direct descendants of Abraham, left in the far west to pursue their own thoughts and enquiries.

Stukeley's elaborate fantasy has, in a number of reincarnations, affected popular images of Stonehenge ever since.

SO WHAT WAS THE PURPOSE OF STONEHENGE?

Nobody knows for certain why Stonehenge was built or what it was used for. Followers of 'New Age' beliefs link it to Earth energies and ley lines – imaginary straight lines linking sacred prehistoric sites – and sometimes to cosmic energy from outer space. Seriously mathematically minded people try to find alignments with the rising and setting points of various celestial objects. Others see it as having been used to predict eclipses.

There certainly appears to have been some astronomical alignment. The main axis of Stonehenge lies roughly north-east to south-west, so that in one direction it points to the sun's rising position in midsummer, and in the other to its setting position in midwinter. However, the case for multiple and detailed alignments has not gained widespread acceptance among archaeologists: there are too many uncertainties in both data and interpretation.

Modern archaeologists mostly hold the view that Stonehenge was built not at a single time but over a period of about 1500 years – from around 3100 BCE to 1600 BCE – and by a number of different cultures.

Development was not particularly systematic: structures appear to have been started and then abandoned and new structures built, or different uses made of the same materials. At one stage the direction of the main axis was changed somewhat. It is likely, therefore, that the views of the builders also changed and evolved over the course of time.

THE STORY OF THE STONES

Ten thousand years ago the last Ice Age was drawing to a close. As the earth warmed, ice sheets melted, sea levels rose and England and Ireland became islands. Open forests of pine and hazel became dark, dense, deciduous forests of linden, oak and elm threaded with fern and ivy. The gently sloping chalk plains of Wiltshire in southern England became open grasslands.

Stone-age hunter-gatherers who inhabited this territory dug a row of pits and in them erected pine poles. The location of these ancient pits is now the car park at Stonehenge. Traces of pine or charcoal in the holes have been dated to the Mesolithic era, 8500 to 7550 BCE: perhaps before Stonehenge there was a Woodhenge.

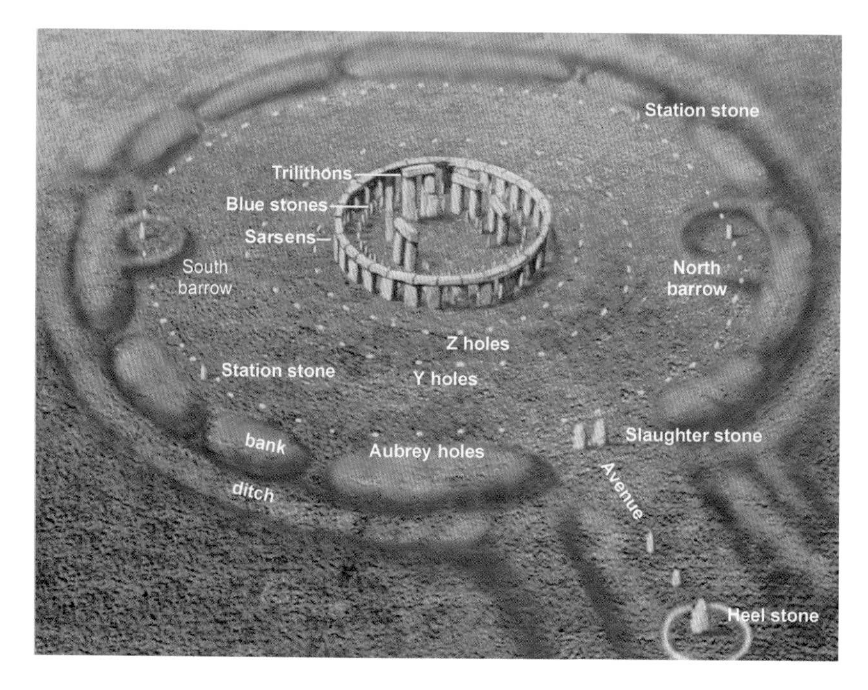

A depiction of how Stonehenge may have looked at its completion around 1600 BCE.

As the climate warmed, more people moved into the area. Six thousand years ago, Neolithic-era farmers are believed to have migrated from the Mediterranean region, displacing the earliest inhabitants, and settling on the coastal plains and easily drained soils of the upland. The heaviest settlement took place on the chalk hills of the south and west.

These new people lived in stable village communities. They cleared the land, planted wheat and barley, raised herds of domesticated cattle, sheep and pigs, used stone axes to clear the forests, and buried their dead in distinctive earth or stone structures known as long barrows. It is now believed it was the descendants of these people who began building Stonehenge.

Deer roamed freely in the forests bordering the Salisbury Plains. As the animals shed their antlers the Neolithic farmers collected them to use as picks and rakes. These tools were used to build the first structures. Teams of people laboured to dig a circular ditch, 100 metres in diameter. The bright white chalk that was excavated was then piled in a long mound around the inside of the circle.

Inside this bank, a circle of 56 holes was excavated and a wooden post placed in each. These cavities, first discovered by John Aubrey, have come to be known as the Aubrey holes. Their purpose remains obscure. Viewed from the centre of the circle, they could have been marker points along the horizon, which would have made them a wonderful tool for studying the movement of the sun, moon and stars. However in the next building phase – from 2900 to 2400 BCE – the posts were removed and the holes partially filled with chalk.

The entrance to West Kennet Long Barrow near Avebury. The main structure was built of stone, which was then covered with earth to form a mound.

Silbury Hill, near Avebury in Wiltshire. This astonishing prehistoric mound is thought to have been constructed in layered steps, much like the Egyptian pyramids. Excavations, starting in 1776, have revealed various small treasures – such as shells, ox bones and antler fragments – but no human bones.

Photo: Diego Meozzi, Stone Pages

The builders

Five thousand years ago, while the ancient Egyptians were building the great pyramids of Giza on the banks of the Nile, people in many different parts of Europe were erecting stone circles, henges, dolmens (simple burial chambers marked by upright stone slabs with a large capstone) and barrows (burial mounds, sometimes of earth, sometimes of stone).

Because of historical accounts, and the lists of rulers painted on the tombs and carved on statues, we know how long it took to build the pyramids. The Great Pyramid at Giza, for example, took around 20 years. But no such records exist for Stonehenge. Archaeological finds, however, suggest that not only did the project take more than 1,500 years to complete, but it was built by a succession of different cultures. It's also clear from the scale that the building work must have involved whole communities. It has been estimated that it took something like thirty million hours of human labour to build the main structures.

Such work could only have been accomplished by a large and successful community, and indeed the land around Stonehenge was farmed intensively and supported a large population.

The early Neolithic agriculturalists who began the construction of Stonehenge were eventually displaced by another wave of migrants. The Beaker People had strong spiritual beliefs and burial customs. Like the Egyptians, they built stone tombs for their most important people, furnishing them with goods – including beakers as drinking vessels – which they would need in the next life. The rest of the dead were buried in single round barrows, with their heads placed at the south end. Men faced east and women west.

As well as farming, the Beaker People made pottery, jewellery and weapons, and probably traded with other cultures all over the ancient world. A later group of migrants, the Battle Axe People, domesticated the horse, used wheeled carts, and smelted and worked copper. They also buried their dead in single graves, often under round barrows, and may have introduced Indo-European languages, of which only Welsh survives today.

Over the years the mingling of these different cultures produced a distinctive group of people, collectively known as Wessex Culture, who dominated the area from 1800 to 1500 BCE. These people were responsible for constructing Silbury Hill, an astonishing mound 39 metres high, built as a series of circular platforms and similar in mass to the oldest Egyptian pyramids. They were also responsible for other monolithic sites in the Stonehenge area – Durrington Walls, Woodhenge and The Sanctuary at Avebury.

WHAT IS A HENGE?

The word henge means 'hanging stone'. Archaeologically speaking, a henge is a roughly circular or oval-shaped flat area enclosed and delineated by a boundary earthwork, usually a ditch with an external bank. The soil and bedrock taken from the ditch were used to build the henge bank, which generally lay outside the ditch.

According to archaeologist Mike Pitts, author of *Hengeworld*, Stonehenge is strictly speaking not a henge as its ditch is outside rather than inside. However every henge is unique: they were not built to a common blueprint.

Access to the central area of a henge was via formal entrances through the earthwork; most henges had either one entrance, or two opposite each other.

Raising the stones

On the site of Stonehenge the builders' attention next shifted from the outer ring – the 100-metre-wide circular ditch – to the heart of the henge. It is possible the first construction was a round wooden building, or woodhenge. The first stone structure, which came next, was composed of blue stones, each weighing about four tonnes. These were brought 220 kilometres by sea and over land from the Preseli Mountains in Wales.

How was this achieved without wheeled transport? One theory is that they were carried by barge to Milford Haven, then along the coastline to the Severn River, and then by barge down the Avon River past Bristol. From there they would have to have been borne across land to the Wylie River, and finally barged up a branch of the Avon to Stonehenge.

The large, familiar sarsen stones with lintels were erected later, from 2850 to 2480 BCE. First the circle of blue stones was removed, leaving just the chalk-filled holes. Next the sarsen stones, each weighing as much as forty tonnes, were dragged 30 kilometres from the Marlborough Downs. Once they arrived they were shaped with round hammers made of stone – no easy task since sarsen stone is very hard.

The sarsen stone circle, 30 metres in diameter, consisted of 30 trimmed blocks topped with stone lintels. Each upright stone was 2 metres wide, 1 metre deep and 4 metres tall. Joints were fashioned to lock the uprights into the lintels, and the lintels into each other end to end. This made a very strong structure. Woodworking technology was employed: the blue stones had tongue-and-groove sides, while the sarsens with matching lintels sported mortice and tennan joins.

Finally, between 2440 and 2100 BCE, the taller, horseshoe-shaped arrangement of trilithons was constructed. How was each huge stone raised and made to stand upright? First a hole would have been dug and the stone positioned with one end over it. The stone would then have been hauled and levered up, while at the same time its lower end was dropped into the hole.

Stonehenge's circle of sarsen stones, with the ditch in the foreground.

Raising the lintel stones was an even more astonishing feat of engineering. The trilithons were seven metres tall, and a lintel stone could weigh as much as nine tonnes. Each would have been positioned on supports next to the uprights and a wooden platform built underneath it. A succession of layers would then have been added to the platform. When it reached the height of the tops of the uprights, the lintel would have been levered into position. The construction was so precise that still today, even though the ground is slightly sloping, the tops of the lintels are almost perfectly level.

By the end of this phase, the trilithon horseshoe, the sarsen circle, the surrounding ditch, the causeway with heel stones, and one station stone on either side were in place.

Finishing touches

Meanwhile work began again on the blue stones. Over the next 250 years a blue stone oval was added within the trilithon horseshoe, and a blue stone circle between the horseshoe and sarsen circle. The blue stone oval was no sooner finished than it was reworked again, this time into a horseshoe. In the final building stage – from 2640 to 1520 BCE – two circles of holes, now known as the Y and Z holes, were dug. Apparently these were for stones to be placed in, but this was never done. Finally the so-called altar stones, made of Cosheston sandstone from south-west Wales, were put in position.

The Stonehenge trilithons. Raised four millennia ago, each stone could weigh as much as fifty tonnes and a lintel nine tonnes.

Four thousand years later, many of Stonehenge's uprights and lintels are still in place. Even more stones might remain were it not for the fact that for many centuries there was little interest in the henge, and nothing to stop local people using it as a convenient source of building material for houses and roads. Still, it is interesting to consider how many structures built today will still be standing 4000 years from now.

STONE CIRCLES AND HENGES AROUND THE WORLD

Monolithic, mysterious stone structures, often in the form of circles and henges, are found in many parts of the world. Some were built as long ago as 6000 years. These structures appear to have been very important to the communities which built them: vast amounts of time, energy, care and ingenuity were clearly taken in their construction. Many have survived not only through many centuries but through many different cultures. Today, archaeologists and scientists continue to try to unlock their meanings and purposes.

CEREMONIAL STONE CIRCLES

Some large and significant structures seem to have been erected for mainly ceremonial purposes – that is, there appear to be no clear astronomical alignments of the stones, nor any burial mounds within or around the circle. Such structures include Avebury, located a few kilometres from Stonehenge, the huge megalithic complex Stanton Drew near Bristol, England, and the Ring of Brodgar in Orkney, Scotland.

Avebury, a great ring of sarsen stone slabs, is surrounded by a massive ditch and a 2-kilometre-long circular bank which was originally 17 metres high. The sarsen stones enclose two smaller circles, each the same size as Stonehenge. Both were probably the focus of ritual and ceremony.

At the **Ring of Brodgar**, 36 stones quarried from solid sandstone bedrock still stand. The ring is enclosed by a ditch over 3 metres deep and 9 metres wide.

Some English stone circles contain remains of wooden structures similar to one believed to have once stood within the sarsen circle at Stonehenge. At the great circle of **Stanton Drew**, which is twice as large as Stonehenge, a highly elaborate pattern of buried pits which once held massive posts is arranged in concentric rings, and there is evidence of additional pits at the centre of the circle.

RIGHT: *Part of the great circle of sarsen stones at Avebury, England. The ditch and bank can be seen in the background. Avebury is the world's largest stone circle, with a 2-kilometre-long bank and 98 stone slabs.*

FAR RIGHT: *One of the great monoliths at Avebury.*

STONES AND GRAVES

Mên-an-Tol in Cornwall, England, believed to have been built in the Bronze Age, up to 4000 years ago. In later times, people crawled through the hole in the belief this would cure conditions such as rickets, rheumatism and back complaints. Photo: Diego Meozzi, Stone Pages

Some stone circles visible today are the skeletal remains of ancient structures. Because so much is missing, their original function can easily be misinterpreted. For instance, although the **Duloe Stone Ring** and the **Mên-an-Tol**, 'stone with a hole' – both in Cornwall, England – resemble free-standing stone circles, it is now believed they were in fact the retaining walls of earthen grave mounds (barrows) or tombs. It is known that in the Bronze Age holed stones were sometimes used as entrances to burial chambers.

The **Bryn Cader Faner Circle** in Wales combines a stone circle with a burial mound. The ruins consist of a small cairn, about 8 metres wide and less than 1 metre tall, with 15 thin slabs, which lean out of the mass of the monument so that it looks like a crown of thorns.

At **Kenmare** ('The Shruberries') in Ireland's County Kerry, and **Carn Llechart** near Swansea in Wales, the stone circles once enclosed central burial mounds with massive capstones. The Kenmare capstone still remains – 2 metres long, nearly 2 metres wide, a metre thick and weighing almost seven tonnes. Kenmare is located near a deposit of copper, a valued resource in the Bronze Age. There are also Bronze Age cairns near Carn Llechart but the 25-stone circle has lost its capstone.

Rollright — Hrolla-landriht, the land belonging to Hrolla — lies in an area of henges. Rollright is a complex consisting of **The King's Men**, a perfect late-Neolithic circle of 77 stones; **The King Stone**, a standing stone 73 metres to the north-east; and **The Whispering Knights**, a 5000-year-old burial chamber marked by five

The beautiful Bryn Cader Faner Circle in Wales may have originally been a burial site. It was pillaged over a century ago, but a hole in the middle suggests the presence of a grave. Photo: Diego Meozzi, Stone Pages

CHANGING ROLES

Just as a building today may have many uses over its lifetime, at some sites ancient stone circles have, over the centuries, found new roles. At **Loanhead of Daviot** in Aberdeenshire, Scotland, for example, the original circle is believed to have been a device for observing the cycles of the moon, useful knowledge for planting crops. Later it enclosed a low cairn and a mortuary pit. Later still, after a cremation cemetery was established next to it, a dwelling appears to have been built on the site. Now only the hearth remains.

Photo: Diego Meozzi, Stone Pages

upright stones leaning together conspiratorially. According to legend, the Rollright stones were once human beings, the army of a king. More probably, however, The King Stone marked the burial mound of a chief who commanded an army.

Stone circles marking burial places are not found only in Britain. A world away, in Gambia and Senegal in Africa, there are also hundreds of stone circles, believed to date from around 750 CE. Most have between 10 and 24 stones, with the largest, at Gambia's N'jai Kunda, weighing about 10 tonnes each. Most are like round pillars with flat tops but there are also distinctive V shapes – believed to mean that two family members died at the same time and were buried together.

This stone circle in Wassu, Gambia, is typical of those found in parts of West Africa. They are thought to be grave markers built around the burial mounds of kings and chiefs. Today, small stones and vegetables are still left on the stones as offerings. Photo: Michael J. West

Set on a mountain pass more than 2000 metres above sea level, the ancient Piccolo San Bernardo stone circle may have been built aligned to the summer solstice. The ring shows through the snow for just a few weeks of the year.

Photo: Diego Meozzi, Stone Pages

CIRCLES MARKING PASSES AND BORDERS

Some stone circles, such as Europe's **Piccolo San Bernado** (or **Petit Saint Bernard** as it is known in France), mark borders and mountain passes for traders and travellers. This large circle, 72 metres in diameter, is located on a mountain pass between Italy and France and straddles the border between the two countries.

Similarly, in the south-west of England **The Hurlers**, three large, aligned stone circles, stand in a pass between two rivers, a good meeting place for traders. Legend has it that the stones were once men, turned to stone for 'hurling the ball' on the sabbath.

CIRCLES WITH ASTRONOMICAL OR CARDINAL ALIGNMENTS

The astronomical alignments of some stone circles, or their alignment with compass points, are readily apparent. These include Long Meg and Her Daughters and Swinside, both in Cumbria, Nine Stones in Dorset, Drombeg (the Druid's Altar) in Ireland, Callanish I in Scotland, Er Lannic in France and Fossa in Italy.

Long Meg and Her Daughters' circle of around 60 stones and Long Meg herself, a 3.6-metre-tall block of red sandstone, are both aligned to the midwinter sunset. In **Nine Stones**, a small, elliptical circle, the two largest stones are placed either side of true north, while the two smallest are directly opposite.

Swinside has a well-defined entrance at the south-east, with two portal stones outside the circumference marking the midwinter sunrise. The tallest stone in the circle stands almost exactly due north. At **Drombeg**, the circle stones slope upwards to a long recumbent stone. The entrance is in the north-east, aligned to the midsummer sunrise. A portal stone is placed so that, when lined up with the recumbent stone, it points to the winter solstice sun and a notch in the distant hills.

The alignments at **Callanish I** in Scotland's Outer Hebrides are lunar. The circle, which dates from around 1800 BCE, consists of 13 tall narrow stones, with another even taller stone at the centre. From this, four incomplete avenues lead away. Looking along the stone avenue to the south, the eye is directed to the setting of the full moon in midsummer.

On the small island of **Er Lannic** in Brittany, France, two 5000-year-old stone circles, which would have once been entirely on land, now lie half-submerged by the waters of the Gulf of Morbihan. The cardinal points of north, south, east and west are clearly marked, and some of the stones probably relate to the setting points of the moon. At **Fossa** in Abruzzo, Italy, large upright slabs are arranged in circles and straight lines, with the latter arranged east to west, probably also indicating astronomical alignment. This site is believed to date from the early Iron Age, 1000 to 800 BCE.

Callanish I, on the Isle of Lewis, Outer Hebrides, Scotland. According to local tradition, when giants of old refused to convert to Christianity, St Kieran turned them to stone. Photo: Diego Meozzi, Stone Pages

OTHER ANCIENT MONUMENTS WITH ASTRONOMICAL ALIGNMENTS

Ahu Akivi, Easter Island's only sea-facing moai, or stone statues.

Photo: Jan Stromme, Lonely Planet Images

Many other sky watchers and shamans of the ancient world recorded the sky's changing patterns in myths and stories, buildings and markers. In the classic era of **Rapa Nui** (Easter Island) culture from around 800 to 1680 CE, nearly 900 huge moai or stone statues were carved, with just under 300 being placed on top of ceremonial platforms known as ahu. One group of seven moai at Ahu Akivi has a precise north-south alignment. They are the only statues on the island that look seaward. All other moai look inland.

In North America the **Chaco** or **Anasazi** civilisation flourished for several centuries from 900 CE, finally vanishing about 1300 CE. The Chaco are believed to have constructed the 'light dagger' in New Mexico's Chaco canyon. At noon at the equinoxes and solstices, a vertical shaft of light passes between two leaning stones and cuts through two spirals carved into the canyon wall.

At **Machu Picchu**, the Inca Temple of the Sun, a carved rock pillar called the Intiwatana – 'hitching post of the sun' – had four corners oriented toward the four cardinal points. Accomplished astronomers, the Inca used the angles of the pillar to predict the solstices, which were critical for agriculture.

Sunlight and shadow playing on the Intiwatana create ever-changing patterns of light and shadow. At a solstice these slow down, stop, and then begin again with new patterns for the second half-year. This performance repeats year after year.

In Mesoamerica, the **Mayan** civilisation flourished between 2600 BCE and 250 CE. The Mayans had a complex belief system based on accurate naked-eye astronomical observations and extensive use of mathematics. Their astronomer-astrologers were obsessed with calculating time, and developed calendars that could predict eclipses with amazing accuracy.

The construction of **El Castillo** – the great pyramid of Kukulcán near Cancun in Mexico – reflects this obsession with time. The pyramid was built to represent Snake Mountain, where, according to Mayan tradition, creation took place. At sunrise, the nine pyramid levels cast a shadow on the staircase edge, producing an undulating snake body climbing the stairs.

Each side of the pyramid is made up of nine tiers. Staircases in the centre of each side lead to the temple at the top. There are 91 steps on each staircase, making a total of 364. With the addition of the step leading to the platform at the top, there are 365, one for each day of the solar year.

Each side of the pyramid has 52 rectangular panels, corresponding to the 52 years of the Mayan cycle of creation and destruction. The stairways divide the tiers on each side into two sets of 9, for a total of 18, corresponding to the 18 months in the Mayan calendar. At solstices and equinoxes, El Castillo casts predictable triangular shadows. The patterns of light and dark triangles form a changing pattern of seven light triangles and six darker shadows, reflecting Mayan belief in a universe composed of 13 compartments.

The Mayan pyramid, El Castillo. At sunrise, shadows appear to cast the image of a snake.

Photo: John Elk III, Lonely Planet Images

RECOMMENDED READING

Astronomy

How to Gaze at the Southern Stars, Richard Hall: Awa Press, 2004

Maori astronomy

Work of the Gods, Kay Leather and Richard Hall: Viking Seven Seas, 2004

Matariki: The Maori New Year, Libby Hakaraia: Reed Publishing, 2004

Stonehenge and other stone circles and henges

Hengeworld, Mike Pitts: Arrow Books, 2001

Stonehenge Complete, Christopher Chippindale: Thames and Hudson, 1994

Acknowledgements

The authors and publishers gratefully acknowledge the following sources: *Hengeworld* by Mike Pitts (Arrow Books, 2001); *Stonehenge Complete* by Christopher Chippindale (Thames and Hudson, 1994); *Stone Pages* (www.stonepages.com); *Megalith Portal* (www.megalithic.co.uk), *Stone Circles of the Gambia* (http://home3.inet.tele.dk/mcamara/stones.html); *El Castillo* (www.isourcecom.com/maya/cities/chichenitza/il.htm).

First edition published in 2005 by **Awa Press** with the **Stonehenge Aotearoa Astronomy Centre**

Written by Richard Hall, Kay Leather and Geoffrey Dobson, with assistance from Robert Adam, Lesley Hall and James Walmsley

Artwork and diagrams by Richard Hall, assisted by Robert Adam and Kay Leather

Photography by Chris Picking, with additional photographs by Sacha Hall, Graham Palmer and agencies as credited

Edited by Mary Varnham, Awa Press

Production by Sarah Bennett, Awa Press

Designed by Anna Brown & Sarah Maxey

Printed in New Zealand by Astra Print

ISBN 0-9582538-7-0
A CIP record for this title is available from the National Library of New Zealand

PHOENIX ASTRONOMICAL SOCIETY

The Phoenix Astronomical Society is an active, dynamic, non-profit-making society of amateur and professional astronomers based in the lower North Island of New Zealand, with members throughout the country. It is affiliated to the Royal Society of New Zealand, Wellington Branch.

The society is dedicated to making astronomy accessible to everyone, whatever their level of interest and knowledge, and members range from absolute beginners through to students, educators, telescope-makers, and experts in optics, electronics and various areas of research astronomy.

Meetings

The society holds monthly meetings in Wellington and Wairarapa, and bi-monthly meetings in Napier.

Newsletter

The society produces a monthly newsletter that provides members with society news and the latest information on astronomical events and discoveries.

Observatory and clubrooms

The Phoenix Astronomical Society has a modern, well-equipped observatory and clubrooms located adjacent to Stonehenge Aotearoa, within easy access of members in Wellington, Wairarapa, Manawatu and Hawke's Bay.

How to join

The Phoenix Astronomical Society welcomes new members of all ages and levels of interest. To join or for further information, visit our website www.astronomynz.org.nz. Or write to Phoenix Astronomical Society, PO Box 2217, Wellington, New Zealand.

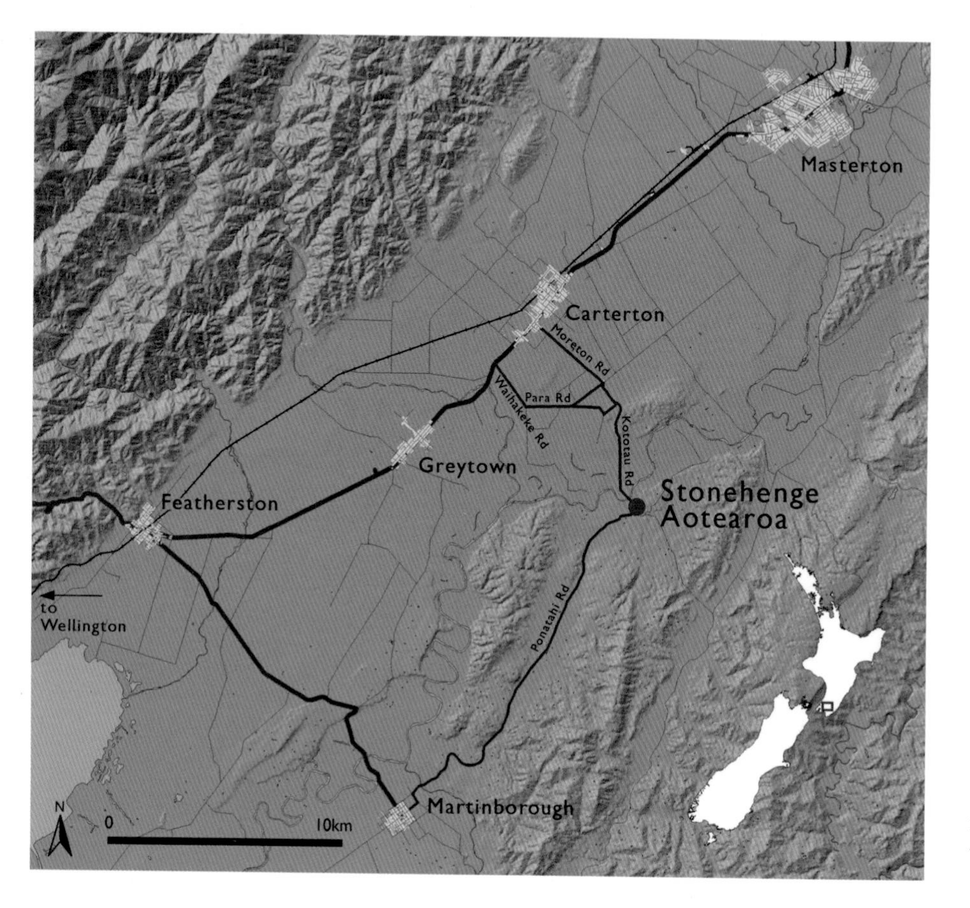

HOW TO GET THERE

Stonehenge Aotearoa is a short drive from New Zealand's capital city Wellington, and from the nearby Wairarapa towns of Martinborough and Carterton.

From Wellington, take State Highway 2 through the Hutt Valley and over the Kaitoke and Rimutaka hills. Continue on towards Carterton, and then follow this map.

Stonehenge Aotearoa
FN 15, Ahiaruhe Road, RD2,
Carterton, New Zealand
Postal address: PO Box 2217,
Wellington, New Zealand
Telephone (within New Zealand):
027-246-6766 or 027-230-5191
Telephone (international):
(0064) 27-246-6766 or
(0064) 27-230-5191
info@stonehenge-aotearoa.com
www.stonehenge-aotearoa.com